Top 40 Costly Mistakes
Solar Newbies Make

*Your Smart Guide to Solar Powered Home
and Business*

**By Lacho Pop, MSE
and Dimi Avram, MSE**

Disclaimer Notice

The authors of this e-book, named "**Top 40 Costly Mistakes Solar Newbies Make: Your Smart Guide to Solar Powered Home and Business 2016 Edition**," hereinafter referred to as the 'Book,' make no representation or warranties with respect to the accuracy, applicability, fitness or completeness of the contents of the Book. The information contained in the Book is strictly for educational purposes.

Summaries, strategies, tips, and tricks are only recommendations by the authors, and the reading of the Book does not guarantee that readers' results will exactly match the authors' results.

The authors of the Book have made all reasonable efforts to provide current and accurate information for the readers of the Book, and the authors shall not be held liable for any unintentional errors or omissions that may be found.

The Book is not intended to replace or substitute any advice from a qualified technician, solar installer or any other professional or advisor, nor should it be construed as legal or professional advice, and the authors explicitly disclaim any responsibilities for such use.

The installation of solar power systems requires certain professional background qualification and certification for working with high voltages and currents dangerous to human life and for installing solar power systems and appliances. The reader should consult every step of your project or installation with a qualified solar professional, installer or technician and local authorities.

The authors shall in no event be held liable to any party for any direct, indirect, punitive, special, incidental or other consequential damages arising directly or indirectly from any use of this Book, which is provided on "as is, where is" basis, and without warranties.

About the Authors

Lacho Pop, MSE, has more than 15 years of experience in market research, technological research and design, and implementation of various sophisticated electronic and telecommunication systems. His extensive experience helps him present the complex world of solar energy in a manner that is both practical and easily understood by a broad audience.

Dimi Avram, MSE, has more than 10 years of experience in engineering of electrical and electronic equipment. He has specialized in testing electronic equipment and performing techno-economic evaluation of various kinds of electric systems. His excellent presentation skills help him explain even the most complex stuff to anybody interested.

You may contact the authors by visiting the website: solarpanelsvenue.com or by emailing them at author@solarpanelsvenue.com

Also by the Authors:

The Truth About Solar Panels: The Book That Solar Manufacturers, Vendors, Installers And DIY Scammers Don't Want You To Read [Kindle and Paperback Edition]

ASIN: B00Q95UZU0, ISBN: 978-6197258011

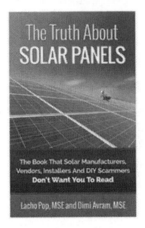

The New Simple And Practical Solar Component Guide [Kindle Edition] **ASIN:** B00TR7IJPU

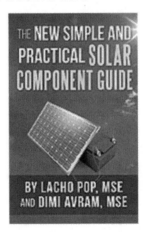

The Ultimate Solar Power Design Guide: Less Theory More Practice [Kindle and Paperback Edition]

ASIN: B0102RCNOG, ISBN-13: 978-6197258042, ISBN-10: 6197258048

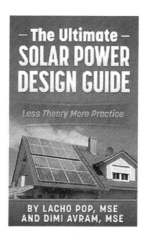

Solar Power Demystified: The Beginners Guide To Solar Power, Energy Independence And Lower Bills [Kindle Edition]

ASIN: B00UC9WWAK

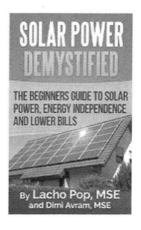

Table of Contents

Introduction

Going solar is considered cool nowadays and, most importantly, can save you money, regardless of whether you live in a city or in a remote area, whether you are connected to a local power distribution grid or not.

Photovoltaics are really becoming more and more attractive. With the demand for solar, there's no wonder that there are thousands of solar-related products available on the market. Fortunately, solar is not rocket science, so you do not have to pay thousands to experts to figure out whether your house is suitable for installing a solar electric system or not. Also, you don't need a solar guru to advise you on what kind of a solar electric system would best meet your daily energy needs.

With the enormous supply of solar related products on the market, however – all the solar vendors, installers, books, leaflets, how-to-do guides, magazines and online forums – there are lots of traps you can easily fall into. This is why it is crucial to have some idea about the basics of photovoltaics, as well as tips on how to assess whether to go solar or not, and hints on how to select the right solar vendor for your case.

In this book, a group of six mistakes is described, which are quite common, no matter whether you have decided to buy a solar electric system or to implement it yourself:

- General mistakes and misconceptions,
- Mistakes during location assessment,
- Mistakes with solar panels,

- Mistakes in solar electric system sizing,
- Mistakes in assembling the system components,
- Mistakes in buying a PV system.

On the following pages, you are going to get familiar with all of these typical mistakes. You can find all of the main types of solar electric systems briefly reviewed at the end of this book.

Important Warning:

Some of the mistakes mentioned herein are actually not targeted towards beginners, since grasping or correcting those mistakes requires practice and a knowledge base that can be acquired in our other books:

-The Truth About Solar Panels: The Book That Solar Manufacturers, Vendors, Installers And DIY Scammers Don't Want You To Read [Kindle and Paperback Edition] **ASIN: B00Q95UZU0, ISBN: 978-6197258011**

-The New Simple And Practical Solar Component Guide [Kindle Edition] **ASIN:** B00TR7IJPU

- The Ultimate Solar Power Design Guide: Less Theory More Practice [Kindle and Paperback Edition]

ASIN: B0102RCNOG, ISBN-13: 978-6197258042, ISBN-10: 6197258048

On the other hand, we really want to help you save money by getting an inexpensive yet effective system built with reduced hardware costs; one that is both easy to maintain and less prone to failures.

We therefore take the risk of not being fully understood by keeping the language quite technical at times.

Above all, unlike some other authors, we want to be honest to our readers and voice our concerns about some of the most frequently committed mistakes that drain your pocket.

GENERAL MISTAKES AND MISCONCEPTIONS

1) Totally 'forgetting' the drawbacks of photovoltaics

If you're planning to buy a solar electric system, please mind the following:

- Electricity generated by a PV system is still more expensive than electricity supplied from utility grid, unless you live in a remote region where connecting to a utility grid would cost you a fortune. PV systems do make solar electricity more affordable (than, for example, it was 20-30 years ago) but prices still remain relatively high. Nevertheless, for the last few years, prices of solar photovoltaic panels have dropped 80% on average and they still continue to decrease.

- Using PV systems for heating is not recommended. For heating, you should use a solar thermal system. Another option for heating is propane or natural gas.

- The high costs of a PV system are concentrated in a substantial initial investment. Often the biggest problem is how to find initial financing. Once a PV system is installed, with its payback spread over a long enough period of time, it is nice to feel independent from the utility grid or to see your monthly electricity bills going down. Buying a PV

3

system is actually like paying your electricity bills in advance for years ahead, and the point is just to avoid the essential burden of high initial costs. That is why it is important to find a suitable source of financing.

- PV systems only produce power when the sun is shining. Therefore, something should be done with the electricity produced – it should either be consumed right away, exported to the grid (in grid-tied systems), or stored in a battery for a later use (in stand-alone systems).

- For people who are connected to the grid, the decision to purchase a PV system is usually based on economics – the idea of reducing monthly bills by selling power to the local utility. For people living in remote areas, who are far from a utility company, the decision to purchase a PV system is not determined by economic reasons, but is rather a matter of securing a normal life instead.

If your home or office is connected to a local utility grid, full replacement of the utility with a PV system might not be cost-effective.

Offsetting a part of your electrical consumption to a solar electric system, however, could be an excellent way to save money on electricity.

The utility company's costs for generating electricity are always lower than yours since any utility spreads the cost for generating electricity among all its customers.

Unfortunately, although solar energy is free, solar equipment is not free. For this reason, being connected to the grid, the price that you pay for electricity is normally lower than the price you pay for solar electricity generated by your own solar electric system.

Thanks to the free solar energy, however, and the savings from monthly electricity bills in a long run, you will not only compensate the hardware and installation cost paid for the solar electric system, but you will also get a positive return on your investment.

Furthermore, home PV systems are usually not practical for powering large heating systems – heaters, huge electric stoves, air conditioners, or electric clothes dryers. Therefore, you should start with improving the energy efficiency of your home.

2) Ignoring energy efficiency

If you are planning to build or buy a photovoltaic system, you should start by increasing the energy efficiency of your building.

Achieving energy efficiency means to reduce your electrical consumption and your monthly electricity bills respectively. Furthermore, making your building energy efficient is a must before implementing a stand-alone solar electric system.

Energy efficiency is important, since saving energy is less expensive than producing energy. By improving the energy efficiency of your building, the cost of the photovoltaic system you are going to install will be reduced.

Reducing electrical consumption can be done in a variety of ways. You should view your home as an energy system comprised of various interrelated parts, with each part contributing to the overall efficiency.

Energy efficiency goes hand in hand with solar, although it does not appear as cool and sexy. To achieve energy efficiency means to reduce the consumption of the devices in your home or office. Poor energy efficiency means a higher solar energy target (due to the high consumption of the devices in your home or office), hence a bulkier and costlier (both in terms of initial investment and maintenance costs) photovoltaic system.

For example, do you know that the average home energy audit finds potential electricity savings of 30%?

The more energy you need in your home or office, the more energy has to be generated by the solar electric system. This means that to meet your daily electricity needs, you have to buy a large and expensive system, in regard to both the initial and maintenance costs. This is valid especially for off-grid systems, most of which are battery-based. It should be noted that cost of a battery bank is between 25% and 50% of the total system cost.

The higher your daily electrical consumption, the larger and costlier the battery array you need, which is not only harder to maintain but also has to be replaced once every five years or more.

In simple terms, achieving energy efficiency means eliminating all the devices comprising a great part of your electricity bill.

Source:
Clean Energy Council, Australia (2008). Electricity from the Sun – Solar PV systems explained, 3rd Edition, June 2008

3) Underestimating the importance of your daily consumption

Whether or not a solar electric system can entirely replace the utility grid and meet your daily energy needs, depends on your daily consumption.

If your home is already connected to the utility grid, completely replacing the utility with a PV system might NOT be cost-effective.

Offsetting a part of your electrical bill through a solar electric system, however, could be the best way to save money on electricity.

Defining your daily consumption (loads) is a vital step regardless of what kind of solar electric system is going to be implemented.

The greater the load, the higher the energy target, and hence the more solar electricity to be generated – which means a 20 to 60% costlier system, both in terms of initial investment and maintenance.

Being unclear about your daily consumption could be a result of two reasons – either neglecting to identify this vital information, an inability to calculate it, or both.

You can reduce your daily consumption by:

- Implementing energy efficiency, thus reducing the price of the solar electric system by 10 to 30%.
- Changing your personal electric use habits by unplugging electrical devices when they are not in use and reducing the time of use. By reducing energy needs, you can find 10 to 30% reduction in price – you might need a 3.5 kWp system of installed power rather than a 5kWp.

4) Choosing a stand-alone system when needing a hybrid one

A stand-alone photovoltaic system generates solar electricity without being connected to electricity grid.

A hybrid system is a stand-alone system with an alternative power source added – wind generator or fuel generator.

Hybrid systems are preferred in cases where high energy consumption and/or long periods of cloudy days require a bulky battery bank, which is expensive both to buy and to maintain.

Choosing a stand-alone or a hybrid system depends on:

- Whether you use the building year-round or seasonally,
- Whether the site is easily accessible or not,
- How much total energy you need daily, and
- What kind of electrical applications you use – whether they are critical or not.

You could do without a backup generator in a stand-alone system but at higher cost – by oversizing your stand-alone PV system and choosing a battery bank with very large capacity.

Such a strategy is highly impractical for two reasons:

- Extremely high initial cost on batteries
- The system will work with maximum performance just a few months a year (probably in winter), while the rest of the time it will work far below its maximum efficiency because it would produce more electricity than you need. Therefore, the value of the electricity produced will probably not be enough to cover the expenses needed for the battery bank maintenance.

5) Trying to install the system by yourself rather than hiring a professional

In many countries, it is required by law that all electrical equipment be installed by a licensed electrician.

On the other hand, many local building and code inspectors do not have enough knowledge of PV systems. This means that even if you follow the rules, you may have problems proving to a code official that you have installed a code-approved solar electric system.

Therefore, we recommended contacting the local building and code officials to obtain any necessary permits and provide them with appropriate information before you purchase and install a photovoltaic system.

A good plan would be to invite them to inspect the installation before the system is completed. This might help your system obtain the needed approval.

Apart from having enough experience, and having undergone the pertinent professional training, professional solar installers will do the work so that your system meets the necessary requirements of all the electrical and county codes.

Installing the equipment might not be that hard, but complying with all of the associated rules and guidelines can be daunting.

Moreover, it is important that during installation process that you protect your roof from damage. An expert installer will ensure that the installation does not damage your roof. Last but not least, professional installers can help you cope with the paperwork and advise you on subsequent property insurance.

MISTAKES DURING LOCATION ASSESSMENT

Performing a site survey is the starting point for every photovoltaic system installation.

When searching for an appropriate site to install a PV system, the following is to be considered:

- Orientation towards the sun
- Lack of any shading obstacles (during the whole day and throughout the whole year!)
- Minimization of the length of the DC cables between the PV array and the inverter
- Aesthetics
- Protection from theft and vandalism
- Easy access for installation and maintenance of the PV array.

Certainly, the greatest mistake is completely neglecting the need for site survey and expecting that a solar vendor will do that for you. Yes, they will... but why not be better prepared to:

- Abandon your solar project due to bad location,
- Learn the performance limits of the system that will be installed at your site,
- Find out how much your solar project will cost, and
- Avoid getting ripped off by a disreputable solar vendor.

11

6) Finding a poor placement site

The PV array should be provided with clear and unobstructed access to sunlight between 9 a.m. and 3 p.m. every day, throughout the year. Mind that even small shadows can severely affect the power output of the PV array.

To achieve the maximum of your shading analysis, you should do a survey during a bright and sunny day, preferably in summer when trees have their full foliage mass.

During the site survey you should be looking for the following obstacles:

- Buildings – be aware of existing buildings and any future possible buildings throwing shade on your PV site
- Chimneys, power lines, poles, hedges and neighboring roofs
- Trees – if you're performing your site survey in winter, remember that in trees look differently in summer than in winter
- Hills and other natural obstacles – mind that in the winter, the sun is much closer to the horizon than in the summer.

A site that is unshaded during part of the day might be partially shaded during other times of the day.

Similarly, if a site is unshaded in the summer, it might be shaded in winter since the winter sun is lower than in summer (and closer to the horizon) and casts longer shadows.

Important:

Here is a summary of what to be careful about while searching for a proper site of the solar array:

- The spot must be unshaded. Try to find a site receiving maximum sunlight throughout the day

- Don't install modules on the ground – place them out of the way of humans and animals.

- Don't install modules near sources of smoke (chimneys) or heat. The latter includes metal roofs as well. If you want to install your solar array on a metal roof, you must provide at least 4 inches (10 cm) spacing between each panel and the roof.

- Check the sun's position during different times of the day. The location should be clear of every object in your yard that casts a shadow – trees, pillars, sheds, etc.

7) Choosing the wrong orientation and tilt of the solar array

A grid-tied (also known as 'grid-direct') system is a solar photovoltaic system that is connected to a utility grid. You can use the solar generated electricity both to meet your daily electricity demands and to export electricity to the grid, for which you get paid.

For grid-direct systems, the orientation and tilt angle of the solar array is usually subject to roof orientation and roof slope.

You should use a compass to check what direction your roof faces, and a spirit level to measure the angle of the roof from the horizontal.

If your site is located in the Northern hemisphere, you should look towards South, East, and West.

If your location is in the Southern hemisphere, you should look towards North, East, and West.

If you live near the equator, you should look towards East and West.

The ideal roof for mounting your PV array is a roof facing South, if you live in the Northern hemisphere, and facing North, if you live in the Southern hemisphere.

Having chosen the right orientation, you have three options for tilting the solar panels, assuming your roof or installation area permits:

- For average yield throughout the year,
- For maximum yield in winter,
- For maximum yield in summer.

Solar energy differs month to month and season to season. This is also true of the sun's position in the sky. That is why you have to choose in advance between the above-mentioned options.

For example, if your solar panels are tilted for maximum production in winter, it means that the chosen tilt ensures solar rays fall almost perpendicularly on solar panels in the winter.

For average yield throughout the year, your solar panels should be tilted to an angle equal to the latitude (in degrees) of your location.

For maximum yield in winter, your solar panels should be tilted to an angle equal to the latitude of your location plus 15 degrees.

For maximum yield in summer, your solar panels should be tilted to an angle equal to the latitude of your location minus 15 degrees.

How to find fast and easy the latitude of your location?

Just go to Wikipedia and search for your location. Then look at the top of the right corner where the location's coordinates are reported. The first left number is latitude of your location, followed by the longitude. If you cannot find your city in Wikipedia, just find the closest big city to it.

Let's imagine that you live in Birmingham, Alabama (USA), and you are curious to find out what the tilt angles should be for the three available options. Now, from Wikipedia we get:

So, the latitude of Birmingham is 33.525°. Therefore, the three solar tilt angles for the three above-described options are as follows:

- For average yield throughout the year: 33.5°
- For maximum yield in winter: 48.5°
- For maximum yield in summer: 18.5°

Maximum power output is achieved from a solar panel perpendicular to the falling sunbeams.

For winter-only applications: you should utilize high tilt angles.
For summer-only applications: low tilt angles are recommended for higher energy output.

Important:

A tip to find the optimal tilt angle of a solar panel:

Place a pencil or a ruler perpendicular to the module surface at solar noon. On a clear day, the object will throw a shadow. If you start adjusting the panel, the optimal position is reached when the shadow disappears.

8) Improperly chosen mounting of the solar array

Below you can find in brief the most common mounting methods.

Pole mounting:

- The best solution for up to 4 solar modules.
- Enables easy adjustment of panel orientation and can be easily accessed.
- To use a pole mount you have to be sure that the site is both secure and clearly visible to the sun.

Ground mounting:

- Does not make sense in case of one or two modules only.
- When you intend to mount your solar array on the ground, be sure to provide a tilt of at least 10°, in order to allow rainwater to easily run off the modules.

Rooftop mounts:

- Safe and secure; however, you might have problems with cleaning solar modules mounted on the roof.
- Solar modules must never be mounted directly on the roof, without leaving some space underneath.
- Solar modules must never be installed on the highest point of the roof since the probability of being hit by a lightning strike increases dramatically.

Pole mounts or ground mounts are preferable to roof mounts.

The PV array mounting type should be selected by carefully considering:

- Orientation towards the sun
- Site shading
- Weather at the location
- Roof material and bearing capacity (in case of roof mounting)
- Soil type and condition (in case of ground mounting).

Important:

Regarding solar array mounting constructions, mind the following:

- Not every mounting construction is suitable for all kinds of modules, as certain kinds of modules are intended for a specific mounting.
- It's a good plan to ask the supplier of the PV modules to install them on the roof.
- To ensure sufficient cooling of the PV modules, enough room should be provided beneath them.
- A design visa and/or a build permit might be required.
- All the necessary construction regulations must be complied with.

9) Considering roof mounting ideal for every case

Nowadays photos with solar panels installed on the roof are really common. Frankly, such buildings do look nice.

Roof mounting, however, is far from the ideal solution, and here is why:

- Roofs are a hazardous place to work. This means that solar modules will be hard both to install and to clean.

- Roofs can easily be damaged in the process.

- Various parts of the roof often are not orientated in the same direction, and this could be a possible source of problems – for more details please, see Section 'Mistakes upon solar panel installation' later in the book.

- Solar modules installed on the roof might not get the maximum radiation not only because the roof slope might be different from the optimal tilt angle, but also because your roof might get shadowed during different parts of the day or of the year.

- Very often solar panel vendors do not offer you solar module mounts. It should be noted that, due to the abundant variety of roofs, it might be a problem to find the proper mounting kit for your roof.

Even if your system will comprise more than two modules, try to avoid roof mounting unless you don't have any other option.

It's also important to locate your solar panels near the battery bank and the charge controller in order to avoid voltage drops and power losses (which can only be compensated by using larger cables that are more expensive), but installing them on the roof might not provide you with such an option.

19

Last but not least, it is important to have your solar panels installed in a place that is far from any possibility of theft or vandalism.

Important:

If you decide to use your roof to install the solar array, you should mind that **access space around the modules usually adds up to 20% of the required area for placing solar modules**!

Don't try to use every last square inch on your roof to install a solar array because:

- The array will be difficult to install.
- The array will be hard to maintain.
- Wind loading at the edge of the roof will increase.
- From a regulatory point of view, you could violate some provisions about providing available space for fire fighters and other personnel that might need to access the roof area.

Consider also the dead spaces around the array. These are the spots that are either shaded or need to be left between the modules.

10) Ignoring the pros and cons of solar tracking

Using solar trackers is another option for squeezing more power from the sun. A solar tracker follows the sun's position and movement in the sky and ensures maximum collection of sun energy by solar panels.

Tracking mounts are NOT economical in case of less than 4 modules.

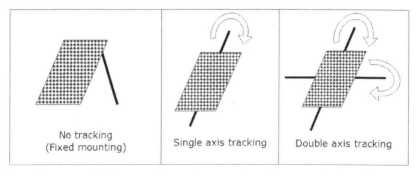

No tracking (Fixed mounting) | Single axis tracking | Double axis tracking

The average efficiency of a solar tracker is reported to increase the total production yield by 25-45%.

Although adding to the overall system costs, residential solar trackers do not need much maintenance. More important is that every solar tracker is a potential point of failure. Furthermore, a solar tracker consumes extra power. Moreover, there might be some local regulations that prohibit the use of solar trackers.

Solar trackers are recommended especially in cases of limited space where customers want to achieve maximum solar array performance.

Source:
http://energyinformative.org/solar-panel-tracking-systems

MISTAKES WITH SOLAR PANELS

11) Mistakes upon choosing the solar panel type

Monocrystalline solar modules are the most efficient solar panels. Their efficiency lies within the range of 12-25%, with typical value of 18%. Use monocrystalline modules if your space is limited or if installing large PV panels would be too expensive.

Although being the most expensive, monocrystalline solar panels have a long proven history of reliable operation. The output of solar panels decreases with every year of operation. However, you can still find monocrystalline solar panels that have been operational for more than 40 years, while still providing about 80% of their rated power. Therefore, in some cases this proven technology can justify the higher price.

Thin-film modules usually need twice the area to produce the same amount of electricity compared to mono- and polycrystalline modules.

Choosing thin-film modules would mean you have both:

- Very limited budget, and
- A large area to install the PV array.

For hot climates and prevailing low solar irradiance levels throughout the year, thin-film solar panels might be the better option. This is because of their better performance under these conditions.

Furthermore, you should have in mind that during the first year of operation thin-film modules produce about 10-15% higher energy. After about six months of their operation, they settle down to their usual yield, a number that they sustain over the rest of their years.

Polycrystalline solar panels currently account for about 55% of the solar installation worldwide, and they are also a good option if you are on a shoestring budget. They are cheaper than monocrystalline solar panels but require more available area to install, compared to monocrystalline ones. In most cases, however, they are the best compromise between price, performance, and available area for installation.

Larger PV modules need less fixing points to the roof thus making installation process less expensive.

Larger modules are more difficult to handle and offer fewer opportunities for various configurations.

All the opposite is valid for smaller modules – they need more fixing points to the roof, are easier to handle, and offer you greater flexibility in creating various configurations on your roof.

The most common PV module size is 1.6 x 0.8 meters, or 5.2 x 2.6 feet.

12) Get tempted by a DIY offer of cheap solar panels

Using secondhand solar panels is not advisable unless you are either on an extremely limited budget, or you need to power a small appliance rather than an entire household.

It's a good idea to evaluate first what rebates and incentives are available to you. In some cases, local and federal benefits might bring down the cost of the total system by 30%.

Therefore, you should choose the preferable option for you:

- To have a brand new solar electric system, with warranty and post warranty support, with insurance, at no risk, no unpleasant surprises, with a lifespan of at least 25 years at a price reduced by 30% than the initially stated one, or

- To have a system built by cheap solar panels – well, probably cheaper than 30% but … lacking any warranty support, insurance, always prone to risks and hazards, of unknown lifespan and most importantly – with no right to benefit from any rebates and incentives.

Here is what you should remember about the option of using secondhand solar panels:

- You are not entitled to receive any grants, rebates or incentives.

- You are not allowed to have your solar electric system connected to the grid.

- If panels are not provided with a certificate, this will be a problem for insurance companies.

- Don't buy any secondhand panels before having a look at them. When checking them out ask: where they have been used, what is their age, and the value of efficiency degrade.

- Buy secondhand panels only when you're either on a very tight budget or when you need them for small electrical appliances (laptops, lamps, fans, water pumps).
- Mind that secondhand solar panels always require larger area to install than new ones.
- Avoid buying thin-film secondhand panels, try to find crystalline (mono- or poly-) ones.
- Seed scams – some 'vendors' might even try to sell you a scam, solar panels manufactured by technology that no one has ever seen!

If you have weighed the pros and cons and have decided to take advantage of a secondhand panels offer, we recommend you do the following before buying them:

- Insist on having a look at the panels,
- Take a fellow electrician, technician or a technology-skilled person with you and ask them to bring a multi-meter along,
- Carefully examine each panel and beware of breaks, chips, scratches, water condensation,
- Have the voltage, power, and current measured,

Having paid a visit to have a look at them, ask the seller:

- How old are the panels?
- In what kind of environment have they been installed (the best case is if they have been stored in a warehouse)?
- How much is the efficiency degrade?
- Is there a warranty offered with the panels?

13) Getting tempted by DIY solar panels

DIY solar panels are said to be the lowest cost option for cheap solar panels. Making solar panels at home is not as easy as described in lots of books and articles.

Homemade panels are NOT recommended to use in solar electric systems with high wattage/voltage/current, including for powering an entire household. Here are the reasons:

- **Shorter lifespan and much faster efficiency degrade** than manufactured solar panels.

  Unless you are able to encapsulate your solar cells effectively, water will get in and the panel performance will degrade over time. This means that your DIY solar panel could have a lifespan of just a few years. Compare this to a typical 25-year lifespan of factory-made panels, combined with a 25-year performance guarantee!

- **Could be a fire hazard** resulting from poor quality soldering, especially when combined with high voltages as a result of connecting several panels in a string. It's also dangerous to make home solar panels using wood and/or plastic.

- **Lack of proper certifications**. This means that:

  o You cannot connect your house to the grid since homemade solar panels are not compliant to the applicable electrical code.

  o You cannot apply for a governmental rebate or incentive.

  o If a DIY solar panel ignites and results in fire damage to your house, your insurance company will not pay an indemnity for a fire caused by a solar panel with no UL certification (for the US).

27

- o Moreover, since DIY panels are not certified, mounting them on any insured structure will void the insurance policy on that structure.

Therefore, DIY panels are not a viable alternative to factory-made solar panels, when it comes to:

- Having a safe and reliable solar electric system
- Cutting your electricity bills
- Qualifying for rebates and government incentives.

Lots of green sites and e-books are full of stuff that seems to work. The problem is, how long will this type of panel last when exposed in the weather?

If you've never done DIY before, and you're now making your first exciting steps into photovoltaics, solar panel assembly is not a great place to learn electrical wiring and soldering.

Source:
Pop, Lacho, Dimi Avram (2014-11-26). The Truth About Solar Panels: The Book That Solar Manufacturers, Vendors, Installers And DIY Scammers Don't Want You To Read The Truth About Solar Panels (Kindle Locations 483-490). Kindle Edition.

14) Mixing different solar panels

Mixing solar panels of various voltage or wattage, or produced by different manufacturers, is an issue discussed by lots of DIY-ers.

Though mixing different solar panels is not recommended, it's not forbidden and things will generally be okay as long as each panel's electrical parameters (voltage, wattage, amps) are carefully considered.

Connecting different solar panels is not recommended because:

- Apart from rated power, each panel has a power degrade percentage. This means that the output of solar panels degrades in a different way over time. Moreover, the stated degradation does not always coincide with what is written on a panel's nameplate. Therefore, it's not easy to find an exact panel match between different solar vendors.

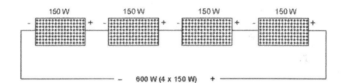

If among the panels connected in series there is a panel with a rated current lower than the others, it will drag down the current passing through all the remaining panels:

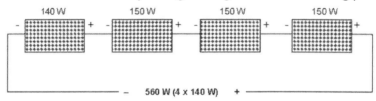

- For panels connected in parallel, the current is additive, while the voltage is the same. If among the panels connected in parallel there is a panel with rated voltage

29

lower than the others, it will drag down the voltage on all the remaining panels.

- Mixing solar panels with different electrical characteristics is not recommended, if you use an MPPT charge controller. Different wattages make it impossible for the controller to find the optimal operating voltage and current, since the voltage and current are different for each panel type.

The solution is simple: utilize panels that have the same or similar electrical characteristics as the original panels.

15) Mistakes upon solar panel installation

Below you will find two common mistakes committed upon solar array installation.

Both mistakes would result from the wish to use the last square inch of the installation area and mount as many panels as possible. Non-observance of some basic rules, however, will lead to degradation in performance.

Being familiar with these mistakes, you will not only be able to avoid committing them if you intend to install the solar array by yourself, but you will also be able to watch whether a solar installer is really doing the best for you.

Sometimes a solar subcontractor will try to convince you that, regardless of the limited area you have, he'll be able to squeeze every inch in order to provide you with the most effective PV system in the world. Okay, you could cool such an enthusiasm and...choose another solar installer.

1) Modules with different orientation connected in a string

If the PV system is based on a single inverter (which is the common situation), the system performance is bound to the performance of the solar modules with the worst position towards the sun. From the above picture, it is obvious that **both** groups of solar panels mounted on this roof cannot get the **maximum** amount of sunlight at a given moment.

An option to avoid this is using an individual inverter for each solar array with orientation different than the remaining arrays.

In such a case, however, the economic viability of installing more than one inverter should be carefully considered in advance.

2) Inter-row shading due to insufficient inter-row spacing

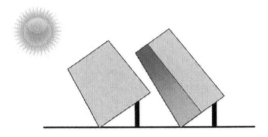

Inter row shading will cause serious degradation of your solar electric system's output, especially if monocrystalline or polycrystalline modules are used.

To avoid inter-row shading, make sure enough spacing is provided between solar panel rows:

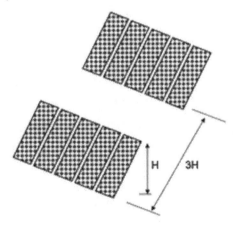

So, to avoid any inter-row shading, you should place two individual neighbor rows at a space, which is at least 3 times the maximum height of a tilted row.

MISTAKES IN SOLAR ELECTRIC SYSTEM SIZING

16) Mistakes upon determining PSH

Peak Sun Hours (PSH) is a measure of the available daily solar resource at your location in hours. PSH is the number of hours representing the total solar energy received on a given day at your location in respect to an irradiation of 1,000 W/m^2. PSH can be easily derived from solar radiation data (also known as 'insolation' data) for a specific geographic location.

Important:

Please, do not mistake Peak Sun Hours with available sunny hours!

For example, 6 sunny hours during a bright sunny day might only translate into 4 PSH per such a day.

PSH value is used to calculate the energy output of a solar electric system.

PSH is determined by solar maps. Solar maps can be used either in a graphical form or as a software database. The latter form is much more precise and provides many more opportunities.

A precise and reliable online source of solar data is posted on NASA's website, the NASA Surface Meteorology and Solar Energy Data Set.

By entering the latitude and longitude of any location on the Earth, you get the daily PSH value averaged both annually and for each month of the year, for various tilt angles, including also the optimal one.

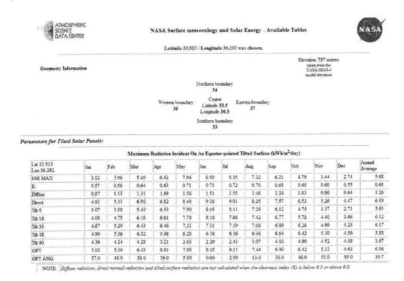

Both the NASA database and graphical solar maps can be used to determine the average annual PSH value, which is used to determine the annual electricity production of a solar electric system.

You can find data about the available PSH at your location by using the NASA source link below.

Source:
https://eosweb.larc.nasa.gov/cgi-bin/sse/grid.cgi

17) Miscalculating the daily consumption

Calculating your daily electricity consumption is a step to both reaching energy efficiency and implementing a solar electric system.

Since an off-grid system is not connected to the grid, it is irrelevant to talk about offsetting a part of your energy consumption to the PV system. The PV system should be able to meet all of your daily energy needs.

Calculating your daily electrical consumption means performing a load analysis – determining your daily electrical energy consumption in Wh (or kWh).

Performing a load analysis is very important since your PV system should be neither oversized (=waste of time and money) nor undersized (=useless for you).

To calculate your total average daily load, you need to determine the amount of energy (in kWh) consumed by each AC load. Therefore, you need to know the rated power of each load, the amount of time it is used each day, and the number of days this device is used each week.

You can find the power rating of each device on its label. If only current (in Amps) is stated, multiply it by the voltage to get the power consumed. You'll get the energy needed for every device by multiplying the power by the number of hours the device is on.

By looking at the list of devices and the energy consumed, you will get the idea about which of them consumes the most energy, and either think about ways for reducing the consumption or discuss some possible alternatives.

18) Misselecting system voltage

For the smallest solar electric systems, a system voltage of 12V is optimal.

A system of such a size is unlikely to be expanded in power output, but it also has the following limitations:

- Cable should be up to 40 feet (120 m).

- Solar generated power can be a maximum of 3,000 watts; this limitation comes from the impractically of high DC current running through the system.

- If you use a 12V system to convert DC into AC power, you won't be able to find an inverter with more than 3,000 watts of power output, which means that you are limited in the number of household devices you can power.

Generally, low-voltage systems are related to higher currents. This means that cable of wider cross section will be required, and this will result in higher costs for cabling.

It should be noted however, that lower voltages pose a smaller electrical risk. Therefore, for the smallest systems, built using short-run cables, 12V system voltage is the optimal solution.

24 volts are ideal for a household solar electric system.

Most companies sell components for this voltage. Moreover, a 24-volt PV system will be easier to expand than a 12-volt one.

With higher voltage, current decreases. This means you could use a smaller cable and hence lower the associated costs for cabling.

24 volts mean longer cable runs, so you have the option to install your solar electric system farther from your house rather than in your backyard or on your roof.

Therefore, with a 24V solar electric system you can maximize your solar power generated output.

Important:

Lower voltages are used with short cables.
Higher voltages are used with longer cables.

Longer cable distances require a 48V system voltage to be used. 48 volts are suitable for very large photovoltaic systems. 48 volts are also suitable if your solar electricity generating system is relatively far from your home.

Furthermore, you have the option to use an inverter of even higher output – 6000W or more.

Higher voltage systems are also a good solution for powering really powerful appliances – large water pumps, air-conditioning systems, etc.

19) Ignoring the advanced system evaluation

'Advanced evaluation' is a techno-economic assessment proving the economic feasibility and sense of buying or building a solar panel system.

Such an evaluation should by all means take into account expected price of grid electricity in the future within the period of guaranteed solar electric system lifecycle, along with any potential incomes from other existing investment options.

The advanced evaluation should provide you with enough data to compare the overall net income of your investment in solar PV system with other existing alternative options to invest your money taking into account:

- Price of solar hardware
- Installation costs
- Annual operational expenses
- Generated 'free' solar energy offsetting these expenses.

By assessing how much money you can save from solar electricity, you can make an informed decision about whether it is worth investing in solar electricity, or whether your money would be better invested in other financial instruments, i.e., bank deposits or other possible investment options you can find at your disposal.

A techno-economic assessment should let you find out:

- The total solar power you need to use and install
- The area you need to install your PV modules and which type of PV modules to choose taking into account:
  o Your solar installation area,
  o Various types of modules available on the market,
  o Your budget.

Once you have chosen your type of PV module, you should determine how many PV modules you need to install and the overall cost of your solar electric system.

Then you should get:

- Your solar energy production costs
- How much you can save by installing a PV system over its guaranteed life cycle
- The payback period of your system.

Although such an assessment is vital, you are actually able to perform it yourself. A techno-economic evaluation will prove the reason for your solar invested money by considering:

- Initial system cost,
- Annual maintenance cost,
- Cost of solar generated electricity,
- The rate of rising grid electricity prices and grid electricity cost savings,
- Costs of getting connected to the grid,
- Solar electric system payback period.

20) Overtrusting free system sizing tools provided by solar manufacturers

Free software for solar electric system sizing can be easily found on the Internet, on sites of most solar vendors. It appears attractive since everybody would be excited to get a solar electric system evaluated for free. At least this is what such free sizing tools claim to do.

By using a free tool for solar sizing, you might end up paying a higher price for equipment you don't need. For example, you might receive a calculation that, in order to meet your daily target, you need 7 solar panels of a specific type by vendor A and 6 panels by vendor B. Every solar vendor wants to have you as their customer, and will act as though no other solar vendors exist. What comes next, you should compare the price of vendor A and vendor B.

Furthermore, these free software tools are not actually what you need as a person trying to decide whether you would benefit from a solar electric system or not. The reason is that you are not provided with any means of economic estimation – that is, evaluate your system with regards to:

- Rising electricity prices,
- Potential annual solar income,
- Annual expenses for solar electric system maintenance,
- Any solar incentives you could be eligible for,
- Any costs associated with getting your building connected, to the grid (in case you live in a remote area),
- The payback period of your system.

Free sizing tools are rather limited. They are not for people who have not already decided to go solar. They are also not for people who need to get a techno-economic evaluation of an optimal solar electric system that would best match their needs, budget, and available roof area. Instead, such free software tools are for people who:

- Have already decided to buy a solar electric system,
- Have already chosen a specific solar vendor,
- Have a substantial knowledge of various solar components offered by that solar vendor,
- Have a high enough budget available.

This could mean that they do not need any economic evaluation of the system, as they are not afraid of rising energy prices.

Furthermore, for those people, it's more important to choose a leading and highly reputable (hence, more expensive) solar vendor than throw a couple of hundred dollars for solar components they actually do not need.

MISTAKES IN ASSEMBLING THE SYSTEM COMPONENTS

21) Believing that the charge controller is useless

If the battery is the heart of a solar electric system, the charge controller is definitely the brain.

What if you don't have a charge controller?

Lacking a charge controller (or charge regulator) exposes the battery to both overcharging and overdischarging, which will eventually lead to a reduction in the life of the battery.

Damage as a result of not having a charge controller is inevitable when you use sealed batteries, since overcharging can easily damage this kind of battery and even cause safety issues.

Should your charge controller be oversized?

It is recommended that the charge controller be oversized by at least 25%, due to the possibility of sporadic increases in current.

When would your system not need a charge controller?

No charge controller is needed for small systems where a 10W or smaller PV module charges a battery of 100 amps-hours of capacity or larger. Such a low-power module is not able to overcharge such a high-capacity battery.

How to select your charge controller?

When selecting a charge controller, the following should be taken into account:

- The system voltage,
- The solar array current (Isc or Imp),
- The battery type.

How the solar array current is related to the charge controller selection is described in the book *The Ultimate Solar Power Design Guide: Less Theory More Practice*.

22) Choosing the wrong type of charge controller

The main charge controller types available today are PWM (Pulse Width Modulation) and MPPT (Maximum Power Point Tracking).

MPPT can deliver 10-30% more energy than the PWM charge controller by tracking and matching the optimal operating point of the solar panel with the battery bank. However, MPPT controllers are more expensive than PWM charge controllers.

What kind of charge controller to choose depends on the specific case and is a tradeoff between getting more power from solar panels and extending battery life.

PWM controllers are less expensive. They are very suitable for small wattage solar electric systems. Furthermore, their efficiency is similar to the MPPT charge controller in hot climates.

Important:

Do you know that an incorrectly selected charge controller can result in losing 50% of the available solar power in an RV system?

This a common mistake usually made with charge controllers by RV owners.

They get a high voltage solar panel at lowest cost per Watt and connect this solar panel or these solar panels to a PWM charge controller, and subsequently lose almost 50% percent of the available solar power.

Why does that happen?

Let's consider a 220 W solar panel with:

- Maximum power point voltage Vmpp =29.1 V
- Maximum power point current Impp =7.56 A

Let's imagine this solar panel connected to a simple RV vehicle solar power system consisting of a solar panel charge controller and a 12V battery bank.

As you know, a PWM charge controller is sized in regard to the current delivered by the solar panels. So, the PWM charge controller will deliver a charging current of 7.56A to a 12V battery bank. If we neglect all the losses of the components of this solar power system, the PWM will deliver only 7.56 x 12V = 90W of power to the battery bank.

In other words, we've lost about 130W of the available solar panel's 220W power!

If we use a Maximum Power Point Tracking (MPPT) charge controller, the current provided to the battery bank will be boosted up to 220W ÷ 12V = 18.3A by such controller.

Such a boost in current is ensured by a current booster, which is an inherent part of every MPPT charge controller. So, in this case, the battery bank will receive 18.3A x 12V = 220W power that could be stored in it.

In an ideal case with no component loses, all solar panel generated power will be stored in the battery bank.

The moral of the story: *In order to minimize power losses when employing PWM charge controller, always connect a solar panel with maximum power point voltage Vmpp voltage closer to the battery bank's voltage.*

The second option is to consider the usage of an MPPT charge controller. Although being the most expensive, its high efficiency will pay off in the long run.

Source:
1. MSE Pop, Lacho, Dimi Avram MSE (2015-02-17), The New Simple and Practical Solar Component Guide (Kindle Locations 1198-1199). Digital Publishing Ltd
2. Department of Energy, Manual for Solar PV Training, 2009.

23) Underestimating batteries

The battery is the most important device in a solar power system because it is responsible for storage and provisioning of electrical power to the loads. Unfortunately, it is also the most problematic device, which goes together with the fact that batteries are a relatively expensive item, whose storage and operation you should be very careful about.

Nevertheless, you don't have to spend much on batteries, if your solar electric system is of a small size.

Batteries that are used in PV systems are suitable for a shallow cycle operation. They are designed to provide a small amount of electricity over a long period of time.

In contrast, the starter battery types used in cars/trucks, are designed to release a relatively large amount of power over a small period of time. For this reason, a car/truck battery is not suitable for a PV system since it will not work at the optimal mode for which it was designed. This translates into a reduced battery capacity, a shorter battery life span, and less stored and delivered power per dollar invested in.

Therefore, while building a solar power system you should try to use a battery designed for a shallow cycle operation, also known as 'deep cycle battery.' However, if you consider using car batteries for your solar power system, please don't forget to adjust the charge controller's battery selector to 'Car battery' setting. This means you should avoid the cheap charge controllers without this setting.

So, avoid using automotive batteries in solar electric systems. If you don't have any other option, then use truck batteries rather than car batteries.

Batteries with lead-calcium plates are less prone to deep discharges than batteries with lead-antimony plates.

Carefully consider the battery size and compare it to the space where you are going to place it. It must not only be large enough but it must also be safe.

If the battery is expected to be moved around rather than stay fixed, you should select a battery type that is resistant to tilting and vibration.

Batteries should be placed in a cool and vented room.

A battery must never be exposed to the sun. If the temperature in the battery room goes above 40°C (104°F), the performance and the life of the battery is going to decline. The room should also be well secured.

To reduce voltage drops, batteries should be placed as near as possible to the solar array and to the charge controller. If you have lead acid battery, however, do not place the charge controller just above it, as the hydrogen gas might cause the controller's terminals and connections to corrode.

24) Selecting an improper battery type

As already said, automotive (car) batteries are not suitable for PV systems since they are not intended for deep discharges.

Sealed batteries are very suitable for **mobile** applications of **small, low-current** stand-alone photovoltaic systems.

Sealed batteries are preferred for a couple of reasons:

- Easy to transport
- Cannot let explosive gases out
- Maintenance-free

There are three types of sealed batteries:

- Liquid ('flooded-cell') batteries
- AGM batteries
- Gel batteries

As a rule, AGM batteries provide a higher current while gel batteries generally have a longer life cycle. Some gel type batteries can be left discharged for a long time without any risk of being damaged, which is impossible for liquid batteries. Gel batteries can, however, be easily damaged by overcharging, while for liquid batteries overcharging is not a great concern.

RV/marine batteries are usually 12V and come with up to 100 amp/hours of capacity. They are not designed for use in frequent charge/discharge cycles typical of home power systems due to their shorter life span.

RV/marine batteries are preferred for small solar power applications, including portable ones, since they are:

- Small,
- Easy to handle,
- Relatively inexpensive.

Golf cart batteries come in higher capacities than RV marine batteries – usually 200 amp/hours and up – and generally as high as 6 volts batteries.

They are intended for cycles of deep discharge, therefore they are a good solution for a small or even medium home solar power system. Although they are more expensive than RV/marine batteries, their price per amp/hours is actually lower.

Industrial batteries are usually lead-acid ones. They come as individual 2-volt units, so you are free to combine them in series to make them fit in a 6V, 12V or 24V solar electric system.

Their capacity varies widely and can be up to 3,000 amp/hours. Industrial batteries have the longest life span and the deepest cycle of charging/recharging.

For these reasons, they are also the most expensive batteries. Since they are also the heaviest ones, and difficult to move and transport, they should be installed in a well-supported place.

Industrial batteries are the best solution for home solar power systems of medium size.

25) Wrongly connecting batteries in the battery bank

Always follow this general rule:

When connecting batteries, do NOT mix batteries of different type, model or age!

Different types and models may range from huge to slightly different characteristics.

This translates into a difference in battery voltage and current, charging states and charging and discharging algorithms that should be employed by charge controller or charge controller block of some modern inverters, which can perform the tasks of a charge controller as well.

Batteries of different type, model, and age usually have different voltage and current. Depending on the type of connection (series or parallel) used to build a battery bank, such mixing could reduce the total battery capacity significantly.

Important:

When connecting batteries in parallel, limit the parallel connection up to 4!

The optimal number of battery strings in parallel is 2. By doing so, you will reduce the loss of capacity in case of a malfunctioning battery in any of the parallel strings.

To connect more than 4 batteries in parallel, please consider using a charge controller with independent current control function per every battery connected in parallel. Thus, you are going to provide optimal charging per battery. Otherwise, the differences in voltage, current, and internal battery resistance will result in suboptimal charging.

26) Neglecting maintenance and operating conditions of batteries

If you have a lead-acid battery bank that offers you the best value for the price, here is how to care for it in the best way:

- Use a charge controller to protect the battery from overcharging and overdischarging.
- Control the battery state of charge by voltmeter and hydrometer.
- Never put anything in the battery except water – the acid does not need to be replaced! Use only distilled water and never tap water.
- Never allow your battery to freeze.
- To ensure maximum battery life, every battery should be kept above 50% state of charge.
- The electrolyte should always be to the indicated level.
- Keep the terminals of the battery and the attached hardware clean.
- Store the battery off the floor – in a wooden box or on a non-metallic tray.
- Avoid storing a lead-acid battery for a long time without recharging it. Sometimes storing a battery without recharging it for even a month can result in the battery not reaching its stated charge capacity or not accepting charge at all!

Important:

A lower depth of discharge (DoD) of the battery, i.e., a shallower cycle operation, prolongs the battery's life span.

For example, a battery can have about 500 charge cycles at 80% DoD, about 800 cycles at 50% DoD and about 1,600 charge cycles at 20% DoD.

Therefore, we should always design our solar battery bank for lower depth of discharge. However, the lower the battery bank depth of discharge, the higher the capacity that should be installed, and the higher the investment to be made.

A battery bank designed for a 20% DoD requires 4 times more capacity than the same bank designed for 80% DoD.

Therefore, the final decision appears as a tradeoff between the available budget and desired battery life in terms of charge cycles available. The most used DoD in contemporary solar electric systems is 50%, followed by DoD of 80%.

If you go for the largest battery capacity, you should mind that sometimes it might turn into a big disadvantage.

Why?

Because when the battery capacity is too high for the installed PV power, the battery might stay partially charged for a long period of time. This will lead to sulfation.

Sulfation is a chemical process of building up lead sulfate crystals due to depriving the battery of a full charge. Unfortunately, it permanently covers the surface of the battery's electrodes. Sulfation is responsible for reduced battery capacity, longer charging times, and shorter battery lifespan. To avoid sulfation in a larger battery bank, we must get it into a full charge state as fast as possible and stay there as long as possible. Since the installed PV power is low, you might use an external power source to charge the battery.

Important:

Battery capacity depends on battery discharge rate.

The capacity of a battery increases at low discharge rates. We can get those discharge rates if we reduce the discharge current.

In practice, this can be achieved by using a lower amount of simultaneous loads. For example, while cooking in a RV vehicle, we avoid using a hair dryer or charging our laptop and all our electronic devices or using any other high wattage device at the same time.

The higher the amount of the devices being used at the same time, the higher the battery discharge current, and the lower the battery capacity you can get.

When you notice that the charge controller disconnects your loads from the battery, please reduce your electrical consumption by half for at least a week, in order to help the battery go faster to fully charge mode.

You will notice when your battery goes into this mode that the indicator of your charge controller is indicating 'Full State.' By following this rule, you'll help your battery stay longer in a 'Full State' charge mode. This will reduce sulfation.

However, if situation becomes frequent and starts bothering you, please consider increasing your installed PV power by adding more solar panels to your system.

Non-optimal battery temperature reduces battery lifecycle.

The optimal operating temperature for a battery is about 20°C (68°F). It allows optimal flow of the chemical processes responsible for storing electrical power, thus ensuring the longest battery life cycle at optimal battery capacity. The battery life reduces by half per every 10°C increase of battery temperature with respect to the optimal 20°C.

For example, if the stated battery lifecycle is 8 years at 20°C, it will reduce to 4 years at 30°C and to 2 years at 40°C. So try to keep your battery bank in a cool place.

Source:

1. Hankins, Mark (2010). Stand-alone Solar Electric Systems. The Earthscan Expert Handbook for Planning, Design and Installation. Earthscan. ISBN 978-1-84407-713-7

2. Department of Energy, Manual for Solar PV Training, 2009.

3. MSE Pop, Lacho, Dimi Avram MSE (2015-02-17), The New Simple and Practical Solar Component Guide (Kindle Locations 480). Digital Publishing Ltd.

27) Selecting an inverter of low quality

Currently, there are three different types of inverters available on the market – sine-wave, quasi (modified) sine-wave, and square-wave.

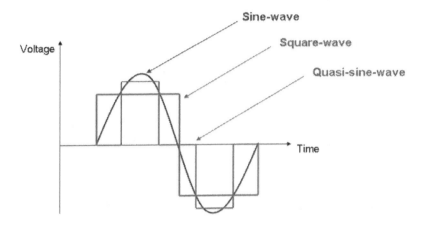

Certain electronic devices, such as mobile phones, microwave ovens, computers, vacuum cleaners, etc., might have problems while operating by a quasi sine-wave inverter.

Furthermore, quasi sine-wave inverters may create additional noise for any audio or TV equipment. Inductive loads such as fridges, pumps, drills, etc., must be powered by pure sine-wave inverters.

A square-wave inverter is of less quality than modified sine-wave inverter.

Although the most expensive ones, **sine-wave inverters are the only possible choice** – not only because they are suitable for any kind of applications, but also because they match regulatory requirements.

28) Confusing grid-tied inverters with off-grid ones

Inverters in stand-alone PV systems are different from inverters in grid-tied systems, although they apparently do the same thing – convert DC into AC electricity.

A stand-alone inverter and a grid-connected inverter cannot be used interchangeably.

The main functions of a grid-tied inverter are:

- Converting DC electricity produced by solar array to AC electricity with sine wave voltage and frequency in compliance to your local standard.
- Disconnecting your solar electric system from the grid during grid power outage to ensure safety of repair work being performed on the grid.

Since every grid-tied inverter stops working during the grid outage, you do not have any electricity during this outage as well. This is so called '**anti-islanding protection.**'

The main function of a stand-alone inverter is converting the output voltage of either the battery bank or the solar array to AC voltage.

In a stand-alone system it is very important that the electricity produced by the PV array be enough to meet the energy needs of the all the electrical loads the PV system is connected to.

Source:
MSE Pop, Lacho, Dimi Avram MSE (2015-02-17), The New Simple and Practical Solar Component Guide (Kindle Location 1389). Digital Publishing Ltd.

29) Improperly connecting the inverter

We use an inverter to get AC voltage (110V-120V/220V-240V) from the DC voltage (12V/24V/48V) produced by a solar panel array connected to a charge controller and a battery bank.

A mistake often made by DIY enthusiasts is connecting the inverter directly to the battery. By doing so, they expose the battery bank to a risk of possible overdischarging.

The inverter should always be connected to the charge controller. In this case, when the battery is running low, the charge controller will disconnect the inverter and the load to prevent the battery from overdischarging.

There is an exception of this rule – you can connect the inverter directly to the battery bank, when you consider the value of the load more important than the battery bank. This means you are ready to pay more for more frequent replacement of your batteries, by regularly and intentionally shortening their lifecycle. Such a situation could happen, for example, when the load is a fridge containing a life-saving medicine, or an important radio device.

If the inverter input DC current is higher than the charge controller DC load rated current, you should connect the inverter directly to battery bank. In such a case, you should have in mind the following.

Important:

When connecting inverter directly to the battery, you can't rely on the inverter's low voltage shut-off function to prevent a battery from overdischarging because its main function is to protect the inverter, not the battery. It is designed exactly for this purpose. Inverter's low voltage shut-off value is too low to prevent a battery from overdischarging.

30) Improperly combining system components

Connections are the veins of every solar electric system.

Here are some common rules you must keep:

- AC loads should be connected to the inverter's output while DC loads should be connected to the charge controller's output.

- Certain appliances, such as low-voltage refrigerators, must be connected directly to the battery.

- In a small DC system with a charge controller, you do not need any fuses other than the one incorporated in the charge controller. In larger DC systems, you need to provide a fuse on the positive terminal of the battery.

- A charge controller should always be mounted close to the battery since precise measurement of the battery voltage is an important part of charge controller's functions. Therefore, even the smallest voltage drops must be avoided.

- A common charge controller has three terminal connections – for the array, for the battery, and for the DC loads. The charge controller disconnects the battery to prevent it from overcharging and disconnects the DC loads connected to the controller 'DC load' terminal to prevent the battery from overdischarging.

- Every device connected directly to the battery instead of the 'DC load' terminal of the charge controller renders the charge controller battery's overdischarching prevention function useless.

- The inverter should be directly connected to the charge controller 'DC load' terminal.

61

- When connecting the inverter to the charge controller 'DC load' terminal, check in the charge controller data sheet whether this terminal is powerful enough to provide the input current to the inverter. Otherwise, connect the higher power inverter directly to the battery bank. In such a case, you will render the charge controller's function that prevents the battery from overdischarging useless.

Another option is to use a second charge controller from the same or another manufacturer, and set it in overdischarging prevention mode. The second charge controller usually sustains high DC currents and is connected directly to the battery for the purpose of load control. On the other hand, the inverter or DC loads are connected to this second controller.

The third option is to find a battery inverter with regulated low voltage disconnect (LVD) that coincides with the LVD parameters of your battery bank. Such an inverter could be connected directly to the battery bank; however, its overdischarging prevention function would be unreliable.

The fourth option is to a use standalone low-voltage DC disconnect device. This device is connected directly to the battery. Then the DC loads or the inverter are again connected directly to this device. You can find devices that support up to 200A DC currents.

Important:

Cheap charge controllers have a low-current 'DC load' terminal. Therefore, their only function is preventing battery from overcharging. You can only connect a low-power 12V lamp or other low-power DC device to this terminal. This terminal switches off to prevent the battery from overdischarging.

In such a case, you must connect the rest of the DC loads directly to the battery. There is no way to disconnect them from the battery in case of overdischarging.

There is a strict sequence to follow upon introducing the charge controller to the solar electric system while connecting and while disconnecting the wires between the solar panel, charge controller, and battery bank:

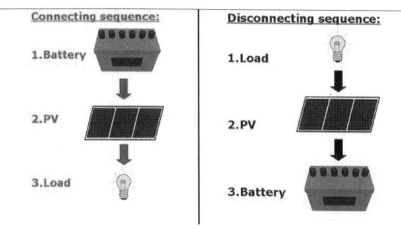

If the battery is not connected to the charge controller *first*, higher solar panel voltage could damage the load!

As you know, the charge controller connects or disconnects the load based on the voltage related to the battery's state of charge.

For example, the solar panel voltage of a solar array for a 12V system may reach up to 18-19V, while the maximum input voltage of 12V-rated device is about 14V.

31) Mistakes in wiring and cable laying

If a solar electric system is regarded as a living organism, wires and cables should be considered as its blood vessels.

We all know how important the cardiovascular system is, so here is some common advice about wiring and cable laying:

- Don't make twisted cable connections as they create relatively high voltage drops.
- Use junction (connector) boxes for all connections.
- When tightening a screw at the connection terminal be careful not to tighten it more than necessary, as the wires can get cut.
- Avoid laying cables that are not purely horizontal or purely vertical.
- Always use a supporting wire when running a cable from one building to another.
- When running a cable along a wall, use clips, at regular intervals, to fix the cable closely to the wall.
- Use a crimping tool whenever possible.

Source:
Hankins, Mark (2010). Stand-alone Solar Electric Systems. The Earthscan Expert Handbook for Planning, Design and Installation. Earthscan. ISBN 978-1-84407-713-7

32) Being not quite aware of voltage drops

Do you know there are certain cases where you don't have to calculate voltage drops?

In the smallest solar electric systems, you can safely make all connections using 2.5 mm^2 wiring cable, as long as the following conditions are met:

- No wire carries current greater than 4 Amps.
- No wire section is longer than 8 m (26 feet).
- Each solar module has a rating of 40Wp or below.

In all the remaining cases, you should consider and calculate voltage drops wherever needed.

Here is a table of suggested maximum permissible cable voltage drops:

Wire run type	Maximum allowable voltage drop
Between battery and charge controller	< 1%
Between battery and inverter	< 1%
Between solar panel and charge controller	< 3%
Between charge controller and loads	< 5%
Between inverter and loads	< 5%

Source:
Hankins, Mark (2010). Stand-alone Solar Electric Systems. The Earthscan Expert Handbook for Planning, Design and Installation. Earthscan. ISBN 978-1-84407-713-7

33) Ignoring system protection

Grounding protection

In most small solar electric systems, all casings of appliances and components (solar array, inverter, and the negative terminal of the battery) should be connected to a main grounding terminal, which is further connected to a grounded electrode.

You should remember the following about grounding:

- Normally solar modules are grounded by connecting to the main grounding terminal. Only when there is a risk of a lightning strike, should the solar array frame be grounded separately.
- Generally, systems with voltages less than 24V need not be grounded.
- If your system contains an inverter and related AC circuits, you should connect all the metal casings and the negative terminal of the battery to a terminal that should be further connected to a grounding rod.

Lightning protection

Important:

Your system NEEDS NOT be protected against lightning strikes, if:

- Your area is not prone to electrical storms, and
- The installed power of your system does not exceed 100 Wp.

Lightning protection will NOT protect your system against DIRECT lightning strikes!

Instead, it protects your system against lightning strikes on nearby objects that are likely to induce high voltages and strong magnetic fields in your equipment and therefore damage it.

Generally, there are two kinds of lightning strike protection:

- External protection systems directing the strike into the ground by means of protective conductors
- Internal protection systems – for example, a DC isolator incorporated into the charge controller, reducing the risk of voltage surges that can easily be transferred to other system equipment.

Overcurrent protection

Overcurrent protection is a must in medium and large home off-grid systems.

The aim of overcurrent protection is to prevent from damaging the circuit wiring, and not the device as a whole.

As a rule, when various wire sizes are available in a circuit, the overcurrent protection should be sized considering the smallest wire in the circuit.

Overcurrent protection is generally implemented by breakers and fuses.

Here is where overcurrent protection needs to be provided:

- Between the solar array and the charge controller
- Between the charge controller and the battery
- Between the battery and the DC loads
- Between the battery bank and the inverter (if any).

Furthermore, it is recommended that you put an overcurrent protection for each DC and AC circuit starting from the DC and AC load center, respectively.

Source:

www.polarpower.org. Photovoltaic Power System (Technology White Paper). Compiled by Tracy Dahl.

MISTAKES IN BUYING A SOLAR ELECTRIC SYSTEM

34) Underestimating the rest of the system costs

Yes, solar panel prices have dropped and continue to drop significantly. However, an operational solar electric system does not consist of solar panels alone. What is more, every piece of equipment should be properly mounted and secured.

Apart from the solar panels, there are other types of equipment such as the other system building blocks (inverter, battery, charger, distribution board), as well as the Balance of System (BoS) equipment – all the cabling, mounting accessories and other type of hardware necessary to successfully launch a solar electric system.

The cost of hardware has not been dropping with the price of solar panels, and this is true of the costs of installation and commissioning labor as well.

35) Ignoring solar panel guarantee terms

One common and rather costly mistake when buying solar panels is not paying attention to the term linear solar performance guarantee.

What does linear performance guarantee mean?

If the manufacturer has stated that a solar panel has linear performance guarantee of 90% for the first 10 years, this means that the panel performance will degrade with the same annual rate within this 10 years period, that is, 1% per year. This is a very good solar panel!

What does it mean if you can't find "linear performance guarantee" in the datasheet and manufacturer has just stated that the panel has only performance guarantee of 90% for the first 10 years?

This could mean that during the end of the first year of operation the performance of a solar panel might drop significantly to 91% of the guaranteed power, and during the next 9 years might only drop with 1% until it reaches the guaranteed 90%.

Have you noticed the big difference? In such a case, for the last 9 years you will incur higher cumulative losses compared to a scenario with a linear power guarantee for 10 years.

Although at first glance there is no any difference between the warranty statement with linear and non-linear guarantee, it is not true.

Source:
MSE Pop, Lacho, Dimi Avram MSE (2014-11-26). The Truth About Solar Panels: The Book That Solar Manufacturers, Vendors, Installers And DIY Scammers Don't Want You To Read The Truth

About Solar Panels (Kindle Location 711). Kindle Edition. Digital Publishing Ltd.

36) Believing that a solar vendor does the best for a solar customer

Suppose you have decided to go solar. You think it is enough to call a solar vendor/installer, and within weeks or months, you'll have your solar dream installed.

Will you be happy however, to find out afterwards that you have paid $15,000 instead of $10,000? Or to discover that your solar investment will not be paid till the end of its lifecycle, as your solar vendor has promised. Solar stuff is business, like many others.

Prior to making a solar vendor promise to do the best for you, you should know what really IS the best for you, in other words, what you actually need. It is really not so difficult to make such an assessment.

Suppose that you call a solar installer and he starts promising you a solar heaven!

Such a solar installer could appear very friendly, convincing, and helpful to you, but he might be far from offering you the most cost-effective solution.

The right approach in such a case is to collect offers from multiple vendors in your area stating to each one of them clearly what kind of system and of what possible configuration you are looking for. Only then could you expect to receive quotes you can reasonably compare (rather than comparing apples to pears).

Such a precondition is necessary since solar vendors are like all the other commercial markets – their target is to sell, as much as possible and at the highest possible price. Your target is to search for the best for you and at the lowest possible price.

To find out what you need for your case, you should get some basic info about various types of solar electric systems and solar panels, as well as to make some simple preliminary calculations.

37) Ignoring any financial incentives available

Here are the most important financial incentives you as a solar user are eligible for:

a) Solar tax credits

People hear about various tax credits related to photovoltaics all the time, but often they do not know how to become eligible for a solar tax credit or even how solar tax credits work. Solar vendors and installers are not tax advisors.

You, as a potential solar electric system owner, should inform yourself about how to take advantage of an attractive solar credit scheme you may be eligible for.

b) Limited time rebates and other incentives

Most of those financial programs are really attractive for everyone who has decided to implement a solar electric system for their homes.

When checking for such incentives available for your area, state or country, you should also look for what period they are offered. Furthermore, you should carefully investigate what requirements need to be met to be eligible for those rebates.

One of the advantages of solar installers is that they are quite well informed about such incentives, and are able to provide you with a good advice.

c) Financing and leasing options

An average photovoltaic system costs between $10,000 and $30,000 after applying rebates and incentives. This is a serious investment, so any option to reduce these costs should be explored.

Most solar vendors will offer you advice to help you get the necessary financing. Another preferred option is a solar lease, which does have its pros and cons.

Source:
Hren, Stephen, and Rebekah Hren. 2010. A Solar Buyer's Guide for the Home and Office, Chelsea Green Publishing, Amazon Kindle Edition.

38) Ignoring the neighborhood environment

Now, an issue like neighborhood character is often neglected because everyone tends to think 'Will it happen to me?' If you live however in a 'bad' neighborhood, you are likely to have your solar panels stolen or find your solar installation shattered by vandals while you are away from home.

The purchase, assembly, and installation of solar panels is always a serious investment. The value of solar panels makes them an attractive target for thieves and vandals.

The reality is that few solar electric systems come with security measures to prevent theft or damage. Therefore, it is often the customer who has to take care of security protection of his solar investment.

39) Misevaluating the pros and cons of free solar panel schemes

'Free solar panels' are actually an offer by a solar power company to install for free a photovoltaic system on your roof and to provide you with 'free' solar electricity:

- **In the US**: against a small monthly fee, thus reducing your monthly electricity bill, or

- **In the UK**: without paying you any monthly fee, with your savings coming from not paying for the part of grid electricity that is replaced by free solar electrical energy.

Here is how the Rent-a-Roof scheme works in the USA:

- A solar power ('Rent-a-Roof') company installs a PV system on your building or home at no cost.

- The rent-a-roof company does all the necessary activities included in the solar electric system launch process – site survey, system design, planning, searching for financing, obtaining permits and installation, and last but not least – performance monitoring during the period of system operation.

- The rent-a-roof company owns the photovoltaic system installed on your house or office building.

- The rent-a-roof company makes a search for all the possible options for qualification for government programs, then does all the necessary paperwork for you.

As a rent-a-roof customer, you have two options available:

- Solar lease – once monthly you get paid a fixed amount by the rent-a-roof company, regardless of how much electricity the solar electric system has produced, or

- Power Purchase Agreement (PPA) – once monthly you get paid for the amount of electricity the solar electric system has produced.

It should be noted that normally you do not have a choice between these two options. Whether you sign a solar lease or a PPA, depends on your location area and on the utility company offering the rent-a-roof scheme.

If you have 'free' solar panels installed on your roof, the rent-a-roof company will buy the electric power generated by the PV system at a rate lower than the utility grid rate.

Actually, the rent-a-roof company, being the owner of the PV system installed, takes advantage of all federal and state subsidies.

Furthermore, the rent-a-roof company commits to buying the generated solar electricity at the lowest annual rate increase compared to the increase rate of electricity provided by your local utility. Therefore, you have electrical bill savings guaranteed in the long run.

You should keep in mind that government subsidies are only paid for a solar electric system that covers annual electricity usage of a family.

So, in terms of money equivalent to electricity exported, if a solar electric system exports more solar electricity than a household consumes, this is not in favor of any rent-a-roof company. Thus, your solar electric system is limited by size. Therefore, you cannot treat free solar panels as a source of increasing income but rather as a way of reducing your electricity bill.

Although free solar panels appear as a win-win offer, there are certain drawbacks. Should you decide to have free solar panels installed, we advise you to consider not only the above-mentioned disadvantages but also to read carefully the fine print in the offered agreement. You should also consider searching out the opinion of a legal advisor.

Sources:

http://www.quora.com/Are-free-solar-panel-schemes-a-good-idea-for-homeowners

http://www.quora.com/Is-an-offer-of-free-solar-panel-installation-from-Peak-Power-US-a-scam

40) Neglecting the relationship between PV and your property

According to a research performed by the National Renewable Energy Laboratory about how your property value is related to having solar panels installed on your roof, each additional $1 in energy bill savings (from your solar installation) adds $20 to your home's total value.

This impressive figure depends on a couple of factors, including your geographical location (areas of active solar markets tend to provide higher return), the size of your solar electric system (the increased property value is directly proportional to the number of solar panels installed), and the value of your home (the larger the house, the higher the property value increase).

Therefore, how much more you get for every $ invested in your solar project varies according to the specific case but recent research shows that the average increase in resale price could be assumed about $6,000 for each 1 kWp of solar installed. This means that 4 kWp installed will add an average of about $24,000 to the retail price of a common house. It should be noted that such assessment is applicable for the currently valid electricity prices. If cost of electricity goes up, the estimated advantage of having solar panels installed will increase as well.

What else can you expect if you have a solar electric system installed, and you decide to sell your house?

A better deal will not only mean better price for you but you could also hope to sell your house faster compared to a house without solar installation. Even if you don't intend to sell, the increased utility rates over the next 20-25-30 years (the lifetime of your solar panel system) will result in savings growth over time.

So, be sure that your solar installation is fully paid for itself.

SOLAR ELECTRIC SYSTEMS AND THEIR COMPONENTS

There are two main types of photovoltaic (solar electric) systems – those connected to the grid and those disconnected from the grid.

Connected to the grid (grid-tied)	Not connected to the grid (off-grid)	
(with or without power backup)	Stand-alone: purely solar	Hybrid: solar with backup generator

Grid-tied systems allow you to offset a part of or all of your electricity demand to photovoltaics, thus reducing your electricity bills.

Most of the existing photovoltaic systems are connected to the local utility grid, so they are 'grid-tied.'

A typical grid-tied (also known as 'grid-connected,' 'grid-direct,' or 'grid-on') PV system generally does not provide electricity storage. This type of PV system generates electricity to provide part of the energy needs of a building in daytime.

Grid-tied systems can be with or without the option for a power backup.

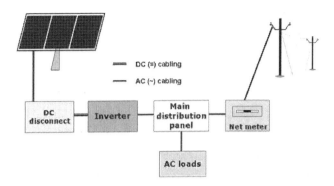

A simplified view of a grid-tied solar system without power backup

Here are the main components of a grid-tied system, which do not provide any power backup:

- Photovoltaic array – generates DC electricity from sunlight
- DC disconnect – disconnects the solar array from the rest of the system
- Inverter – converts DC electricity into AC electricity
- Main distribution panel – the connection point between home electrical network and utility grid
- AC loads – the devices operating on AC electricity
- Net meter – measures the electricity imported from and exported to the utility grid.

The above picture does not show any Balance of System (BoS) equipment – the mounting and wiring systems and components needed to integrate the PV system into the existing building infrastructure. The BoS equipment comprises various cables, jumpers, boxes, protection devices, etc.

When power generated by the PV system exceeds the building's energy needs, the surplus power is exported to the grid. This is called 'net-metering.' It provides you with the opportunity to get paid for the electricity you supply to the grid.

A grid-tied system does not operate during a utility power outage unless it has a power backup.

Grid-tied PV systems without power backup have the following advantages:

- Require almost no maintenance
- Seasonal changes in solar radiation are not as important for their operation
- Are easier to design and are less expensive than stand-alone systems.

In cases where frequent grid outages happen for relatively long periods, a **grid-tied system with power (battery) backup** is recommended:

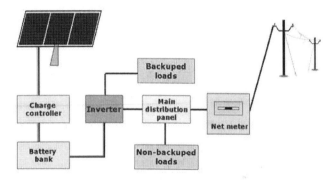

A simplified view of a grid-tied solar system with power backup

The components of such a system are:

- Photovoltaic array – generates DC electricity from sunlight
- Charge controller – regulates battery charging, thus increasing battery lifespan
- Battery bank – stores the electricity generated by the PV array
- Inverter – converts DC electricity into AC electricity
- Main distribution panel – the connection point between the home electrical network and the utility grid
- Backuped loads – all the AC and DC devices provided with power backup

- Non-backuped loads – those electrical devices that are not provided with power backup
- Net meter – measures the electricity imported from and exported to the utility grid.

Another main type of solar electric systems are **off-grid** ones. They can be stand-alone (purely photovoltaic) and hybrid ones.

Off-grid solar electric systems are not connected to the grid.

They are preferred in remote areas where buildings are far from any utility infrastructure. In such a situation, it is cheaper and easier to install a PV system to meet your daily electricity needs rather than pay for utility interconnection.

First, let's describe in brief stand-alone solar electric systems.

The simplest kind of a stand-alone PV system can be obtained by directly connecting a DC load to the PV array:

The load might be, for example, a DC fan or a DC water pump. Such devices use electricity right away after it is generated (i.e., in daytime), without any need to store it for later use.

Off-grid systems are usually provided with battery backup to store the generated energy in case it is not used right away:

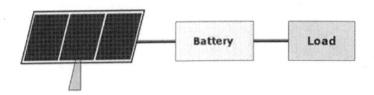

The load might be, for example, a TV set or a laptop. Since such devices operate not only in daytime, a battery is needed to ensure their operation during night hours.

Here is an example of a stand-alone system designed to replace a utility grid for remote buildings:

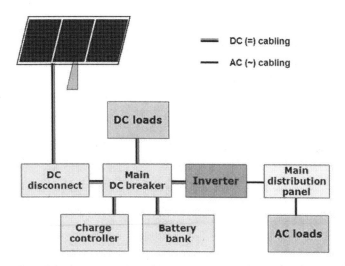

A simplified view of a stand-alone solar electric system

Here is a list of components of a stand-alone system:

- Photovoltaic array – generates DC electricity from sunlight
- DC disconnect – disconnects the solar array from the rest of the system
- Main DC breaker – connects the inverter to the battery and charge controller
- DC loads – all devices operating on DC power
- Charge controller – regulates battery charging, thus increasing battery lifespan
- Battery bank – stores the electricity generated by the PV array
- Inverter – converts DC into AC electricity

- Main distribution panel – the connection point between home electrical network and utility grid
- AC loads – all devices consuming AC power

Stand-alone systems are 'photovoltaic-only' systems. They contain no additional power generator apart from the solar array.

The second subtype of off-grid systems are **hybrid systems**.

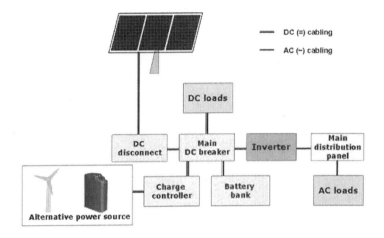

A simplified view of a hybrid off-grid system

A hybrid system is actually a stand-alone solar electric system with an alternative power source added – wind generator or fuel generator.

Hybrid systems are preferred in cases where high energy consumption and/or long periods of cloudy days require a bulky battery bank, which is expensive both to buy and to maintain.

Important:

Whether to choose a stand-alone or a hybrid system depends on:

- How much total daily energy you need;
- What kind of electrical applications you use – whether they are critical or not.

A hybrid system is suitable:

- If daily consumption of electricity is more than 2.5 kWh, or
- For regions with poor sunlight for long periods.

In every solar electric system the solar panel energy generating capabilities are mainly responsible for solar electricity production and correspondingly for solar electricity production cost.

By carefully engineering and designing the rest of the components of a solar electric system, you can squeeze up to 30% more power under the same conditions while keeping the price of your solar installation almost the same.

Furthermore, in the case of an unprofessional system's engineering and design, you may not only lose the additional 30% of energy, but you may lose even more energy at the output of your system. Thus, it would turn out that you have bought the most expensive solar panels on the earth!

How is that possible? Can you really increase the production yield of your solar electric system, even if you have no technical background and you don't know very much about solar power?

Yes, you can. Even if you are not technologically inclined, our informational 'How to' solar packages can help you to quickly build, install, and evaluate an easy and cost-effective solar electric system.

Source:
Mayfield, Ryan. 2010. Photovoltaic Design and Installation for Dummies, Wiley Publishing Inc.

Thank you very much for reading this book!

For updates about new releases, as well as exclusive promotions, visit the authors' website http://solarpanelsvenue.com where you can find many free solar calculators and other useful solar information.

We need your valuable feedback. If, by any chance, you are disappointed with our book, please let us know directly via our personal email: author@solarpanelsvenue.com.

We will work on your feedback and fix our mistakes in our updated version.

If you like this book, we would be more than happy if you would leave a review. Your review does matter! It helps future readers and helps us improve the content of our book. Thank you in advance for this gesture of goodwill.

Also by the Authors:

The Truth About Solar Panels: The Book That Solar Manufacturers, Vendors, Installers And DIY Scammers Don't Want You To Read [Kindle and Paperback Edition]

ASIN: B00Q95UZU0, ISBN: 978-6197258011

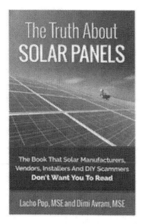

The New Simple And Practical Solar Component Guide [Kindle Edition] **ASIN: B00TR7IJPU**

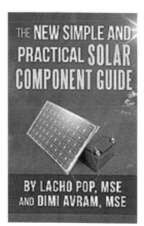

***The Ultimate Solar Power Design Guide: Less Theory
More Practice*** [Kindle and Paperback Edition]

ASIN: B0102RCNOG, ISBN-13: 978-6197258042, ISBN-10: 6197258048

Click on the link to get it NOW:

http://www.amazon.com/Ultimate-Solar-Power-Design-Guide-ebook/dp/B0102RCNOG/

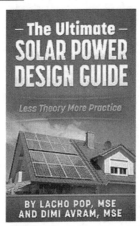

Click on the link to get it NOW:

http://www.amazon.com/Ultimate-Solar-Power-Design-Guide-ebook/dp/B0102RCNOG/

Solar Power Demystified: The Beginners Guide To Solar Power, Energy Independence And Lower Bills [Kindle Edition]

ASIN: B00UC9WWAK

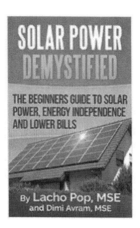

Glossary of Terms

Alternating current (AC) – electrical current changing its direction at a given interval.

Balance of system (BoS) equipment – all the equipment apart from the solar array that is needed for a solar electric system to operate.

Battery – a device capable of producing DC electricity and storing it for later use.

Battery bank – a combination of batteries connected together.

Capacity of a battery – the amount of electricity a battery can store. Capacity is measured in Amperes-hours (Ah).

Charge controller – a device managing the process of battery charge and discharge.

Current – directional movement of electrons upon certain voltage applied.

Conductor – a thing or substance where electric current can occur.

Disconnect (breaker) – an electric switch protecting an electric circuit from overload.

Direct current (DC) – electric current flowing always in the same direction.

Distribution panel (distribution board) – a device dividing electrical power supply into several electrical circuits.

Energy – the work that can be done within a certain period.

Energy efficiency – a set of measures resulting in electrical consumption reductions.

Fuel generator – a generator working on combustive fuel, able to generate AC electricity.

Grid-tied (grid-direct, grid-connected, grid-on) system – a solar electrical system producing electricity that can be both used in your home/office and exported to the grid.

Hybrid system – an off-grid system that combines photovoltaics with additional power sources (e.g., combustive fuel generator, wind generator, etc.) to produce electricity.

Insulator – a thing or substance in which no electric current can flow.

Inverter – a device converting DC into AC electricity.

Load – a device consuming electricity to do some useful work.

Net metering – the process of measuring the solar electricity exported to the grid by a solar electric system owner, credited by the local utility company.

Off-grid (autonomous) system – a solar electrical system disconnected from the grid and producing electricity for home/office use only.

Photovoltaic (solar) cell – the smallest semiconductor unit producing electricity when exposed to sunlight.

Photovoltaic (solar) generator – solar array with all its cabling and disconnects

Photovoltaic (solar) module – a combination of solar cells connected together.

Photovoltaic (solar) panel – a combination of solar modules connected together. Often, although not fully correct, terms 'solar panel' and 'solar module' are used interchangeably.

Photovoltaic (solar) array – a combination of solar panels connected together.

Photovoltaic (solar) system – a combination of solar modules and other equipment connected to produce electricity for practical needs.

Power – the rate of consuming/generating energy.

Semiconductor – a stuff where electric current can only occur under certain conditions. Solar panels are made of semiconductor (silicon) material.

Stand-alone system – an off-grid system that uses solely photovoltaics to produce electricity.

Voltage – a difference in the potential (hidden) energy between two points, causing current to flow upon free electrons available.

INDEX

Made in the USA
Lexington, KY
16 July 2018